CARNIVOROUS PLANTS

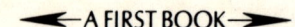

by John F. Waters
illustrated with photographs

Carnivorous Plants

*Franklin Watts, Inc.
New York/1974*

Cover by Michael Horen

Photographs courtesy of:
Charles Phelps Cushing — pp. viii, 12, 50, 51, 52.
Field Museum of Natural History, Chicago — pp. 4, 7, 15. New York Public Library Picture Collection: pp. 11, 18, 19, 20, 23, 26, 28–29, 30, 33, 36, 40, 43.

Library of Congress Cataloging in Publication Data
Waters, John Frederick, 1930-
    Carnivorous plants.

   (A First book)
   SUMMARY: Describes various insect-eating plants including the Venus Flytrap, pitcher plants, and bladderworts.
   Bibliography: p.
   1. Insectivorous plants – Juvenile literature. [1. Insectivorous plants. 2. Plants] I. Title.
QK917.W28   583   73-21976
ISBN 0-531-02700-7

Copyright © 1974 by Franklin Watts, Inc.
Printed in the United States of America
6 5 4 3 2 1

CONTENTS

What Is a Plant?
1

What Is a Carnivorous Plant?
6

Pitcher Plants
14

Bladderworts
25

Sundews
32

The Venus Flytrap
39

Butterworts
45

Animal-Capturing Fungi
49

Glossary
55

For Further Reading
57

Index
59

CARNIVOROUS PLANTS

WHAT IS A PLANT?

Plants are just about everywhere. It is impossible to travel very far on earth and not see some kind of plant. They can be found in lakes and rivers and in shallow waters of the sea. They live where it is very hot, where it is very cold, in the light, and even in the dark. They live in valleys, on mountains, in bogs and marshes, in the desert, and along the beach. Plants practically cover the surface of the earth.

Animals eat plants. Cows eat grass, and birds eat berries and nuts. Insects eat plants. A single caterpillar can destroy a plant all by itself. Bees obtain sweet nectar from flowers so they can make honey.

Plants are very important to people. Some plants are good to eat. Farmers grow fruit trees and nut trees, harvest them, and sell the fruit and nuts in the marketplace. Some farmers grow wheat and corn to sell, as well as tomatoes, cabbages, lettuce, carrots, and many other plants.

Plants are used for decoration. Many people take pride in a rich, green lawn or a handsome bed of flowers. They grow shrubs around their houses and plant green hedges instead of building fences. Trees are often planted near roads and highways to beautify the area.

There are about 350,000 different kinds of plants in the world. They come in many sizes, from the simple one-celled forms

Plants grow almost anywhere. This is a plum tree orchard in bloom in California.

(1)

called *algae* to the huge sequoia trees of California that may grow to be over thirty stories high. Some sequoia trees have been growing for 3,500 years.

Many plants are green because they contain a green pigment called *chlorophyll*. This extremely important agent absorbs light energy from the sun. It changes the light energy into a chemical energy that allows the plants to grow. Using this chemical energy, green plants are able to manufacture their own food materials, so in order to live they must have light. Some plants that have chlorophyll do not look green. They may be brown or red, such as some fresh-water algae and seaweeds.

Most plants that do not have chlorophyll cannot make their own food. A number of these are *parasites* and live on other living plants. Others live on plants that have died and are decaying. One plant that is a parasite is the mushroom. It has a web of very tiny threads that take nourishment mainly from decaying plant matter.

In order for green plants to reproduce, they must flower and produce seeds. Some plants have very tiny flowers; others have large, showy, and colorful blosoms. Flowers need to be pollinated before they can produce seeds that are fertile. When fertile seeds fall to the ground new plants are produced.

Plants produce pollen, which usually looks like fine dust. Male pollen and female eggs must come together to make a seed fertile. Some plants are able to pollinate themselves, whereas others are pollinated by insects. Bats, birds, or even the wind may carry pollen from one flower to others. In cases where wind carries

Colorful, fragrant blossoms, such as the wild rose, attract insects.

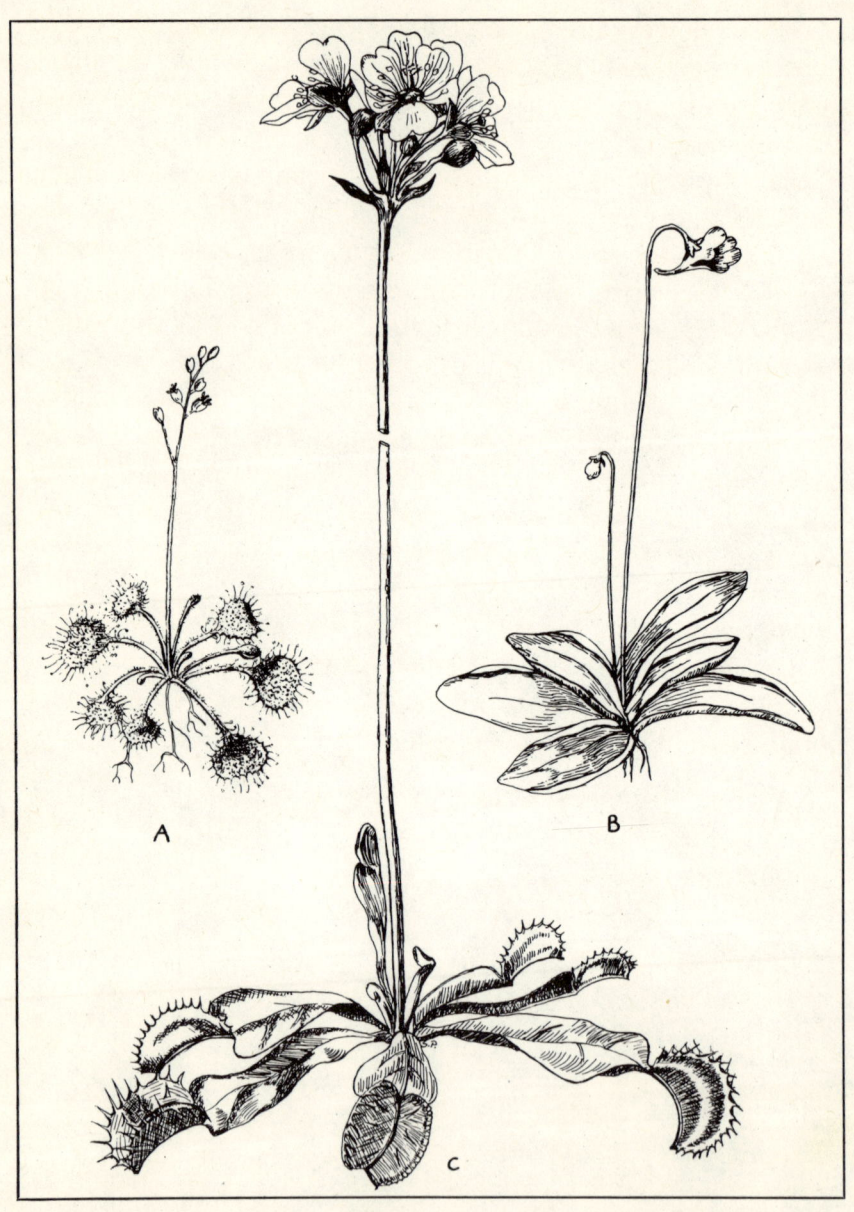

the pollen, flowers are usually not large and attractive. But, where an insect, such as a honey bee, is the pollinator, the plant usually has showy, colorful, and often fragrant flowers that attract the insects.

Generally speaking, most plants have certain basic parts. They have roots that grow down into the ground. Besides giving the plant support, the roots absorb food and water, which is transferred to other parts of the plant. As the water is brought up from the soil, certain chemicals enter the plant with the water. The plant needs these chemicals to help it grow.

Roots are attached to the stem of the plant, also called the shoot. Leaves are attached to the stem, which can be entirely belowground, but is usually mostly aboveground. Inside the stem, water and chemicals are carried from the roots to the leaves and, in turn, food that is produced in the leaves is carried to the roots.

Look around you. No matter where you are, there is a plant or a part of a plant near you. It could be in the form of a wooden door or floor, a piece of furniture, a sheet of paper, or almost anything else. Plants are everywhere.

Most plants have certain basic parts: roots, shown in A; and leaves attached to a stem (shown broken in C).

☙ WHAT IS A CARNIVOROUS PLANT? ☙

Of the thousands of plants in the world, some of the strangest are the ones that are able to obtain some food material from the bodies of insects that get caught in or on some part of the plant. In many cases, these plants have very specialized structures that enable them to trap the insects. Plants in this category are known as *carnivorous* plants. Carnivorous means "that which eats flesh," and, therefore, such plants are often called flesh-eating or insect-eating. But those words can be misleading. While carnivorous plants do "eat" insects, they are not able to behave like animals and actively catch and eat them. In all cases, insects arrive on the plants by chance or are attracted by some scent or color and may become trapped there. In time, parts of these insects are digested and absorbed by the plants. So, "flesh-eating" when speaking of animals is quite different from "flesh-eating" when referring to carnivorous plants.

About 500 different kinds of carnivorous plants are known throughout the world. There are probably ten times as many stories about them. The stories may tell of a huge meat-eating plant that grows on the tropical island of Malaya. If an unsuspecting person ventures near one of these plants, he is grabbed or snapped up and immediately surrounded by a gigantic leaf. A bystander can do nothing but listen to the screams as the victim is eaten alive!

Of course, that story is not true. Of the 500 or so kinds of

Despite stories of plants that eat humans, the meat-eating parts of a plant are too small to consume anything very large.

insect-eating plants in the world, not one eats humans — nor pigs nor horses nor rabbits nor anything else of any size. There is, however, truth in the reports of small rodents falling into *pitcher plants* and not being able to escape. In time their bodies are consumed by the plant. But there is good reason why such plants cannot eat humans. None of the meat-eating parts of the plants is very large. Although some carnivorous plants may consist of vines thirty to forty feet long, their pitchers are still small and delicate.

Even though stories of man-eating plants are untrue, carnivorous plants are nevertheless very interesting. Insect-eating plants grow in bogs, in damp sand, or along the distant edges of lakes and rivers. They also flourish in spring-fed ponds or in the connecting bays of giant lakes, such as those near the Great Lakes. They can also be found in heaths (usually large, level wastelands) or glacial tarns. A tarn is a small pool on a mountain that was carved out by glaciers thousands of years ago.

Insect-eating plants are somewhat fragile, and they cannot stand any kind of upset in the conditions around them. If any type of contamination, such as sewage or industrial waste, oozes into the bog or bay where they live, they will be some of the first plants to die.

All carnivorous plants are very similar to other green plants. They have leaves and stems, but unlike most other plants they lack a large root structure. Their leaves are green, they have chlorophyll, and they are able to obtain energy from the sun. All carnivorous plants can exist without eating flesh. That is, they are normal plants. However, by digesting insects, they do obtain an extra food supply and probably grow better than carnivorous plants that do not have insects as a food supplement.

The shape of the leaves on carnivorous plants differs from ordinary plants. Whereas a leaf of an ordinary green-leaved plant is usually flat, the leaves of the insect-eating plant are often shaped in some special way to catch or hold an insect. In both cases, the leaves take up the energy from the sunlight, but in carnivorous plants, these leaves may also be able to trap food in the form of insects.

Carnivorous plants have flowers and seeds. Their flowers attract insects to aid in pollination.

Insect-eating plants differ in the way they catch or hold their prey. The traps on carnivorous plants can be divided into three general groups:

Plants with pitcher- or bladderlike leaves. The pitcher of the pitcher plant is really a very specialized leaf, and contains a fluid into which insects fall and are drowned and digested. The bladders on the *bladderwort* are also specialized parts of leaves, but these are normally found submerged in water, whereas the typical pitcher plants have pitchers above ground level;

Plants, such as the *Venus Flytrap*, with leaves having two distinct halves that fold over the insect and trap it;

Plants, such as *sundews* and *butterworts*, with leaves that have sticky glands. The insects become stuck on them and are digested.

It is important to remember that carnivorous plants contain chlorophyll and can make their own food by using the energy from sunlight. However, where the soil is deficient in certain elements, especially nitrogen, or where these elements are not readily available, the plants are able to obtain the elements as a result of digesting the bodies of insects and small animals.

Some carnivorous plants, especially the Venus Flytrap, are endangered today. This means that they are no longer in great abundance in their native habitat and are in danger of becoming very scarce or perhaps extinct. People often dig up the plants to keep or to sell as novelties. Most people who buy them treat these plants much as they would regular house plants, and as a result the carnivorous plants die. Such plants should not be removed from the wild. However, anyone who has carnivorous plants should treat them as the special plants they are.

Because of the unusual characteristics of carnivorous plants, they are often brought into the home or the classroom for observation. These are plants that have been grown commercially for just such reasons. All of them must have moist areas in which to live. In the wild, pitcher plants grow in bogs and marshes and live best when some of their roots are in water. Sundews live on moss-covered rocks and logs that are a few inches above the water surface. They do not live directly in the water.

The Venus Flytrap lives in moist soil that has good drainage. The soil cannot be soggy or wet. The air around a meat-eating plant should be moist. A greenhouse is fine, but most schools or homes do not have greenhouses. If anyone put a pitcher plant in a flower pot, the same way as a geranium, the pitcher plant would soon die. No matter how much the plant was watered, it would die due to lack of moisture in the air. Dry indoor air causes the leaves to dry up.

To ensure that the air is moist around a flesh-eating plant, it should be planted in a glass *terrarium*. This can be any large glass jar, provided that it can be covered. A glass cover over a fish tank also works well. The cover holds the air inside, keeping it moist all the time.

An engraving of meat-eating plants, including the pitcher plant (upper left, marked #1), the Venus Flytrap (lower right, #8), a butterwort (lower middle, #7), and a sundew (behind the Venus Flytrap). To the extreme lower left can be seen the tiny balloonlike sacs of the bladderwort.

Preparing a terrarium

To prepare an aquarium or glass jar for meat-eating plants, first cover the bottom of the terrarium with an inch or two of coarse gravel. Then cover the gravel with several inches of acid soil. If acid soil is not available, use ordinary soil found outside and mix it with peat moss. Peat moss can be obtained from any garden shop. Dried sphagnum moss is also good to use, and can be found near bogs and marshes or lakes. The top layer in the terrarium should be entirely sphagnum moss. Plant the carnivorous plant so that the roots are in the soil and covered with both moss and soil. Water should be added — remember that pitcher plants like the soil very wet, sundews like it just soggy, and Venus Flytraps need good drainage.

Place the glass terrarium in a window where there is some good light every day. Do not keep it in direct sunlight for a long time. The sun could cause the temperature inside the glass to rise greatly, and then the leaves of the plant could burn.

Transplanted carnivorous plants may take two weeks or longer to adjust to their new surroundings. Old leaves may die and the plant will look dead, but in time new leaves will grow and the plant will begin to thrive.

New plants can be grown from seeds, but great care is needed. Seeds of the Venus Flytrap, sundews, and pitcher plants may be purchased from biological supply houses.

To grow plants from seeds, use a shallow pan and place one to two inches of gravel on the bottom for drainage. On top of the gravel lay several inches of sandy soil. Then add a thin layer, no more than a quarter-inch deep, of chopped peat moss or chopped sphagnum moss. Sow the seeds in the moss. Keep the pan moist all the time. Moisture can be ensured by covering the pan with a glass cover. Within two to four weeks pitcher plants will begin to sprout, but sundews and Venus Flytraps are irregular — it could take eight days or as long as a month.

🐜 PITCHER PLANTS 🐜

A ladybird beetle, dressed in a bright orange jacket, walked along the ground. Then it crawled up the outside of a strangely shaped plant until it reached the top. The plant looked a little like a milk pitcher. The beetle crawled around the lip of the pitcher, attracted by sweet-smelling nectar. The insect ate the nectar, reaching down beyond the edge of the pitcher. Suddenly, it slipped and fell. The ladybird beetle landed in some liquid. Immediately, it tried to crawl out, but the inside walls of the pitcher had stiff hairs pointing downward and the sides were slick. The beetle tried to fly, but its wing only struck the sides, and it fell back into the liquid. In time the beetle died. Parts of the little animal were digested by the plant. All that remained was the bright orange jacket, its tough *chitinous* (a hard substance) outside skeleton.

That is the typical way in which pitcher plants attract their prey. These insect-eating plants are found in North America from Labrador and Newfoundland to Florida and across the eastern coast to Louisiana. Other kinds of pitcher plants live in Egypt and other parts of Africa, as well as in Australia and the tropics of Asia.

Pitcher plants grow in wet and soggy cedar swamps and bogs, and also in damp fields, grassy swamps, beaver meadows, and

Parts of the pitcher plant: A shows the entire plant, greatly reduced; B shows the pitchers, about one-fourth natural size; C shows part of the flower spike, about one-third natural size.

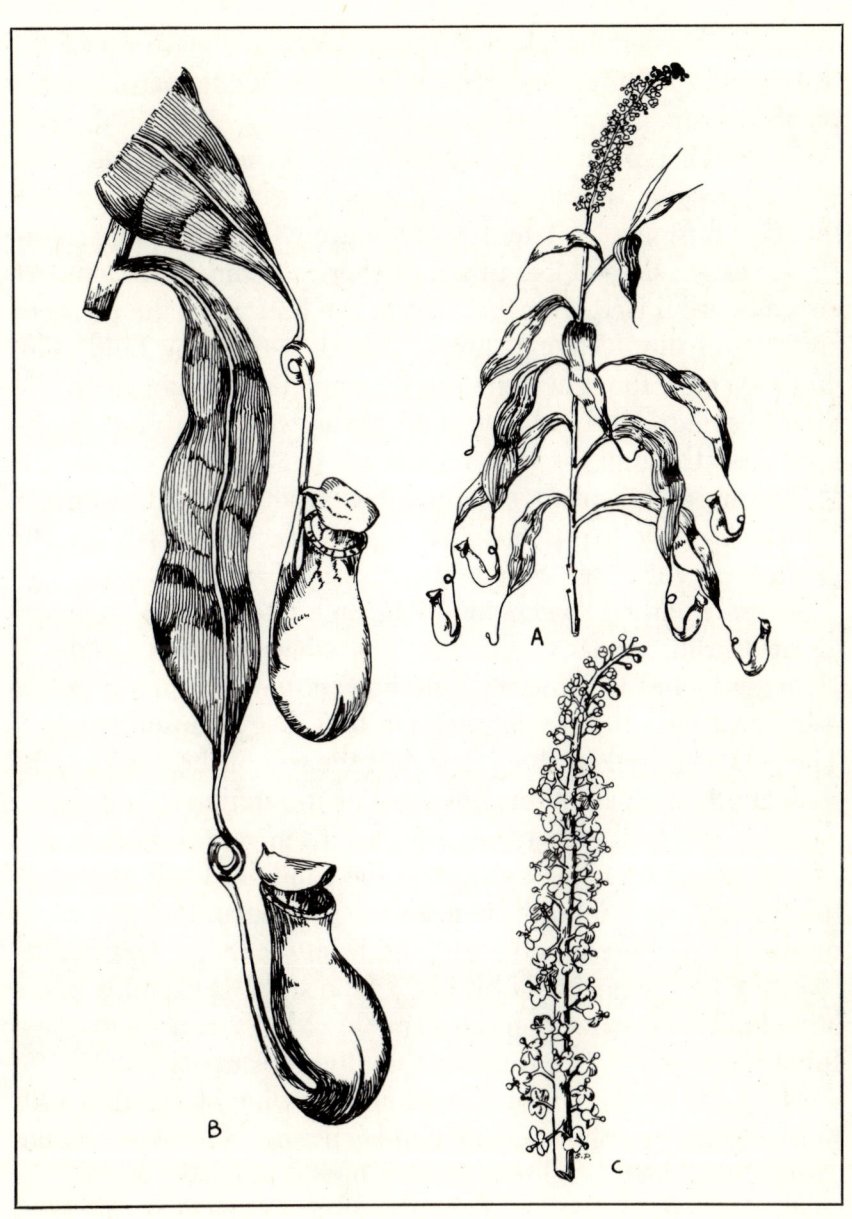

along the borders of lakes. Because of the hollow form of the leaves and the broad lips, they were given the name of pitcher plants.

The pitcher plant is a *perennial*, which means it will grow year after year, as opposed to an *annual*, which grows one season, dies during a frost, and does not grow the following season. Its leaves are shaped like tubes, pitchers, or trumpets. Digestive enzymes are mixed with the fluid at the bottom of the pitchers, and act on the decaying insects that drown in the fluid. This fluid was once thought to be the accumulation of rainwater plus digestive material. However, most of the water is absorbed into the leaves through the root system.

Flowers of the plant are produced from the plant's crown. They each have five petals and five *sepals*. The sepals are the green outer coverings seen on the flower before it blooms.

Several nodding flowers highlight each pitcher after it blooms in late spring or early summer. The stems reach a height of eighteen inches to two feet. The flowers have drooping petals, usually varying from a rich maroon to a deep crimson in color. The petals last only a short time, but the rest of the flower — the stalk and five sepals — remains most of the summer.

When these strange plants are wet from morning dew or a recent rain shower, they glisten in the sunlight. Each pitcher or tubular leaf has a reddish tinge on the outside and is light green inside. The texture is soft and smooth but leathery. Toward the roots the leaf is narrow, a bit like a pipe stem. It expands into a large hollow receptacle and is capable of holding a small glassful of liquid. Even in a drought the pitcher is rarely empty.

Along the inside part of the pitcher is a wing or flap that adds to its strange appearance. The flap keeps the water from evaporating. Small flies, beetles, and other insects enter the pitcher for shelter and its nectar, and then cannot escape. On the sides there are dozens of thick, bristly hairs that grow downward. Anything that attempts to crawl up the sides of the pitcher is hindered by these hairs that line the upper part of the tube and lip.

The common pitcher plant *Sarracenia purpurea* is a widespread species that grows in North America from Labrador to Florida and west to Louisiana. Because of its great attraction, it is the provincial flower of Newfoundland where it carpets bogs for miles and miles. It grows in full sunlight in live sphagnum moss with its roots barely reaching into the soil or muck. This common insect-eater shares the bogs with Labrador tea plants, as well as with cranberries, sundews, some orchids, and sheep laurel. Late May is blooming time, which continues well into July. The leaves can be green, purple, or red in color or tone. The flowers are reddish purple.

In warm climates *Sarracenia purpurea* will flourish throughout the year. In cold climates the leaves drop off when it frosts, but the roots remain alive all winter. When the warming spring arrives the roots produce new leaves. These leaves, which are flat and very thin, push out along the ground. As the leaves grow they begin to widen toward the top, looking much like a giant pea pod.

The leaves actually grow horizontally along the ground with their openings pointing upward. As they continue to grow, deep purple veins begin to show on the outside. In time the leaves reach a length of about six inches. The roots take in water and pass it along to the leaves or pitcher where the tiny reservoir is formed. Cells on the inside of the leaf produce nectar, the sweet-smelling perfume that attracts insects.

While the leaf is forming, the large hairs grow downward. These are the traps that keep victims such as spiders, moths, flies, wasps, bees, beetles, crickets, and ants from escaping.

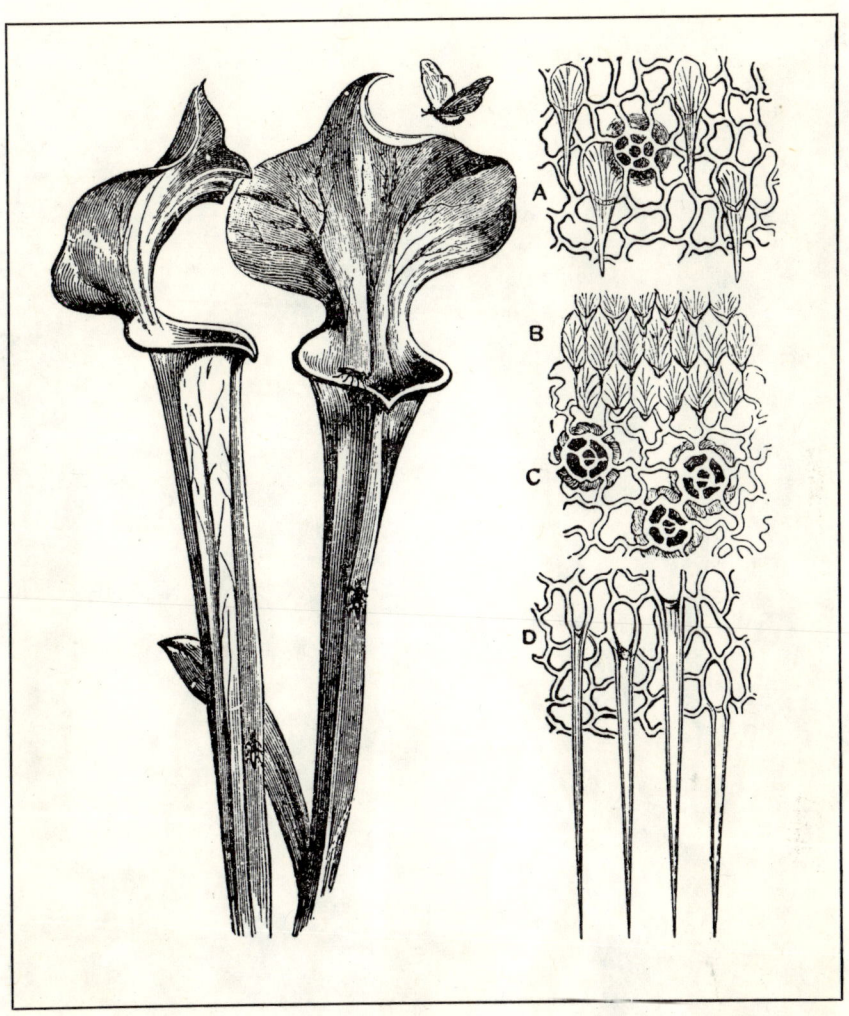

*Left: the common pitcher plant
Sarracenia purpurea
Above: leaves of the
common pitcher plant, showing
magnified surface details*

Left: the hooded pitcher plant
Sarracenia minor
Right: the parrot pitcher plant

Not all insects die in the pitcher plant, however. A moth caterpillar with the delightful name of *Papaipema appassionata* burrows into the plant and feeds on tender parts of the pitcher. For some reason it suffers no ill effect from the fluids. One kind of mosquito lays its eggs in the fluid of the purple pitcher plant. After the eggs hatch, *larvae* grow in the fluid. In time the mosquito reaches maturity and flies away unharmed.

Another pitcher plant, the yellow *Sarracenia flava*, grows along the southern coastal plain of the United States from Virginia to Florida. Its three-foot high, narrow, green trumpetlike leaves and yellow blossoms fill the wet bogs and pinelands. They are in bloom from April to May.

Catesby's Pitcher Plant, *Sarracenia catesbei*, has narrow trumpets, or pitchers, with wide lips. Its flower is pink and green, and the leaves have purple markings. These pitchers differ greatly from those of the hooded pitcher plant, *Sarracenia minor*, which bend over at the top, forming a cover or hood over the opening. The yellow flowering stalks and leaves reach a height of about two feet. This plant grows along the southeastern coastal plain of the United States from North Carolina to northern Florida.

Three pitcher plants are not as common as the others. The sweet pitcher plant, *Sarracenia rubra*, grows from North Carolina to western Florida. It has erect, hooded pitchers about twenty inches high. The flowers may be crimson, purple, red, or green, and they stand taller than the leaves.

The parrot pitcher plant, *Sarracenia psittacina*, has tiny pitchers only six inches high, with hoods that curve over. Its flowers are purple and green with some red, and they grow in pine forests from Georgia to northern Florida and west to Louisiana.

The cobra plant, *Chrysamphora californica*, shares features with the other pitcher plants. Its pitchers grow up to thirty inches tall and curve over somewhat like the head of a cobra snake. Near the opening that points down is an appendage that looks like the "snake's tongue." The plant has purple-red flowers and grows in bogs in northern California and in southern Oregon. It is the only pitcher-producing plant that grows in the far west.

Animals crawl up the outer part of the cobra pitcher plant following a nectar trail. The top has a small hole near the tongue that is really an extension of the leaf. The animal crawls into the small hole still following the nectar. The crown of the cobra plant also has many non-green patches. Light shines through the patches in the same way as it shines through the opening. Because there are so many false openings, the insect is often unable to find the single escape route. In time it loses its grip and falls down into the fluid at the bottom, where it drowns and is eventually digested. Bacteria in the fluid help to break down the body of the animal.

On the island of Borneo in the South Pactific, *Nepenthes tentaculata*, another pitcher plant, grows in abundance high in the trees throughout the moss forest. The striped and gaudy, bluish-purple pitchers dangle from twisted stems. Inside the pitchers, which are about six inches long and two inches wide, are the remains of small flies, beetles, and moths. Some of the pitchers also have mosquito larvae and small fly larvae that are not affected by the digestive powers of the pitcher plant. They live in the water untouched and are able to leave when grown. Spread over these long pitcher plants are glands that secrete a sweet juice. The juice is attractive to colonies of ants, which are lured to their deaths by the hundreds.

A Nepenthes *pitcher plant*

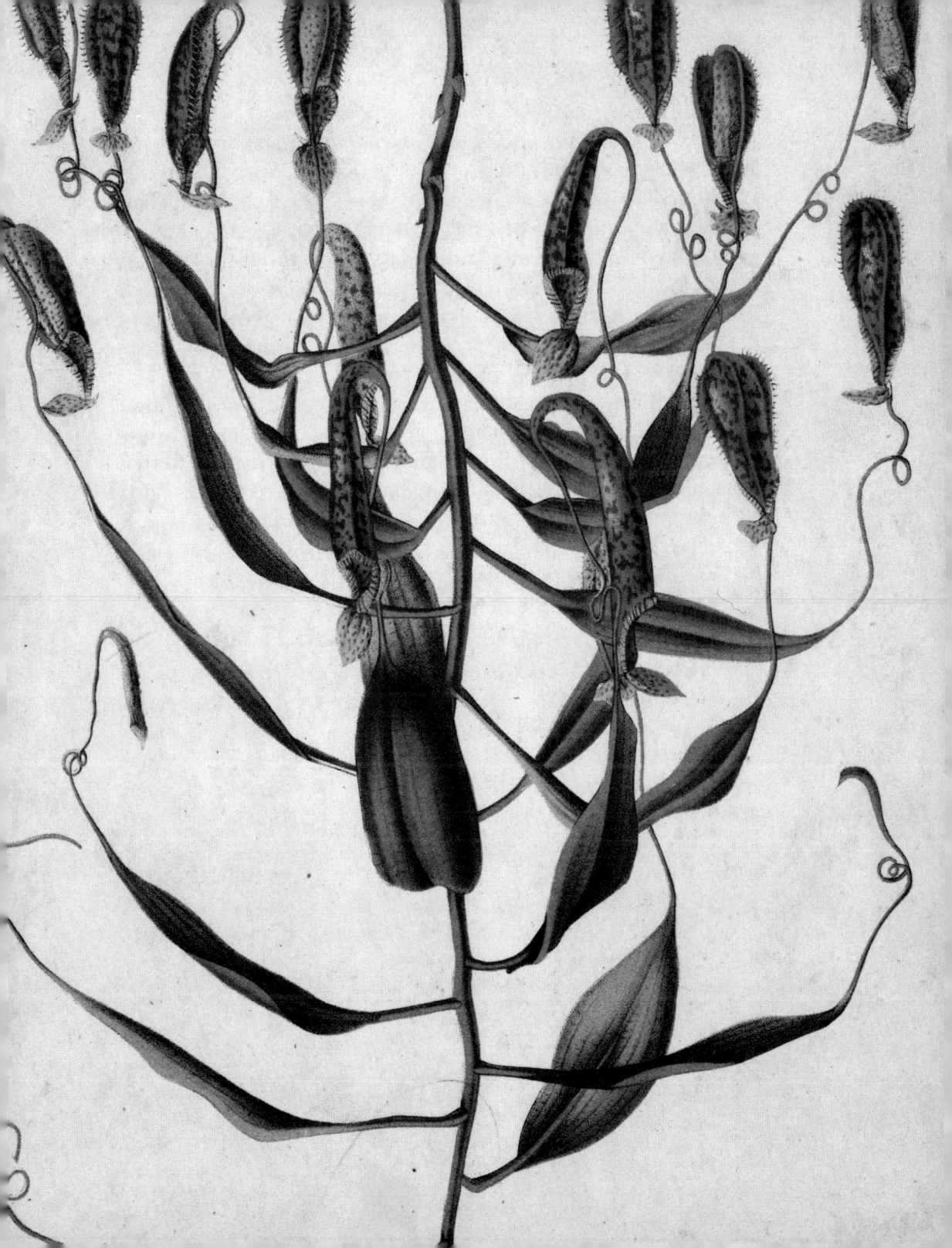

The king of the pitcher plants is *Nepenthes rajah*, first discovered in 1851 on the slopes of Mount Kinabalu on Borneo. It was found at heights ranging from 4,000 to 9,000 feet. This plant grows in the form of a shrub about four feet high, out in the open. The pitchers are squat and fat and reddish-brown in color. Just below its downward pointing spines or hairs are pores that lead to nectar glands. It has a green lid with pink veins. Some of the pitchers grow up to six inches across at the lip, quite large in comparison with pitchers of other plants.

At one time it was thought in various parts of the world that pitcher plants had great medicinal qualities. A concoction was prepared from the root of the plant and given to those suffering from the violent symptoms of smallpox, in the past a dreaded scourge but largely controlled by vaccination today. However, it was later learned that the liquid taken from the pitcher plant had no effect on the disease.

⚜ BLADDERWORTS ⚜

One of the most interesting plants decorating ponds and ditches with colorful flowers is the bladderwort. There are about 275 different kinds of bladderworts throughout the world. They grow on every continent and large island. The water bladderwort, with the scientific name of *Utricularia*, has tiny balloonlike sacs produced on the plant below water or soil level. They are the smallest plant traps known. Insects are trapped in the tiny bladders.

All water bladderworts are similar. They have many stems and leaves. Attached to the leaves are the bladders. Above water is a stalk that bears small flowers less than a half-inch wide. They appear in various colors — blue, yellow, purple, or white, depending on the type.

The common water bladderwort blossoms about mid-summer in the northern sections of the United States. When in bloom the blossoms look striking in full yellow across a pond, peat bog, or sandy shore. Sometimes they are so numerous that they completely cover a bay, pond, or lake, and their yellow blossoms stretch for hundreds of feet across the water during the latter part of July and into the month of August.

Although most bladderworts are water plants, some are land plants, such as one that grows in South America on the forest floor among decaying plants and leaves. This plant has green leaves and green leaf stalks. Food storage organs that look like potatoes are located in the center of the plant. Several colorless stems grow from the area of the organs. Each of these stems carrying the tiny bladders grows in the decaying litter on the floor of the forest. Very tiny animals, so small they cannot be

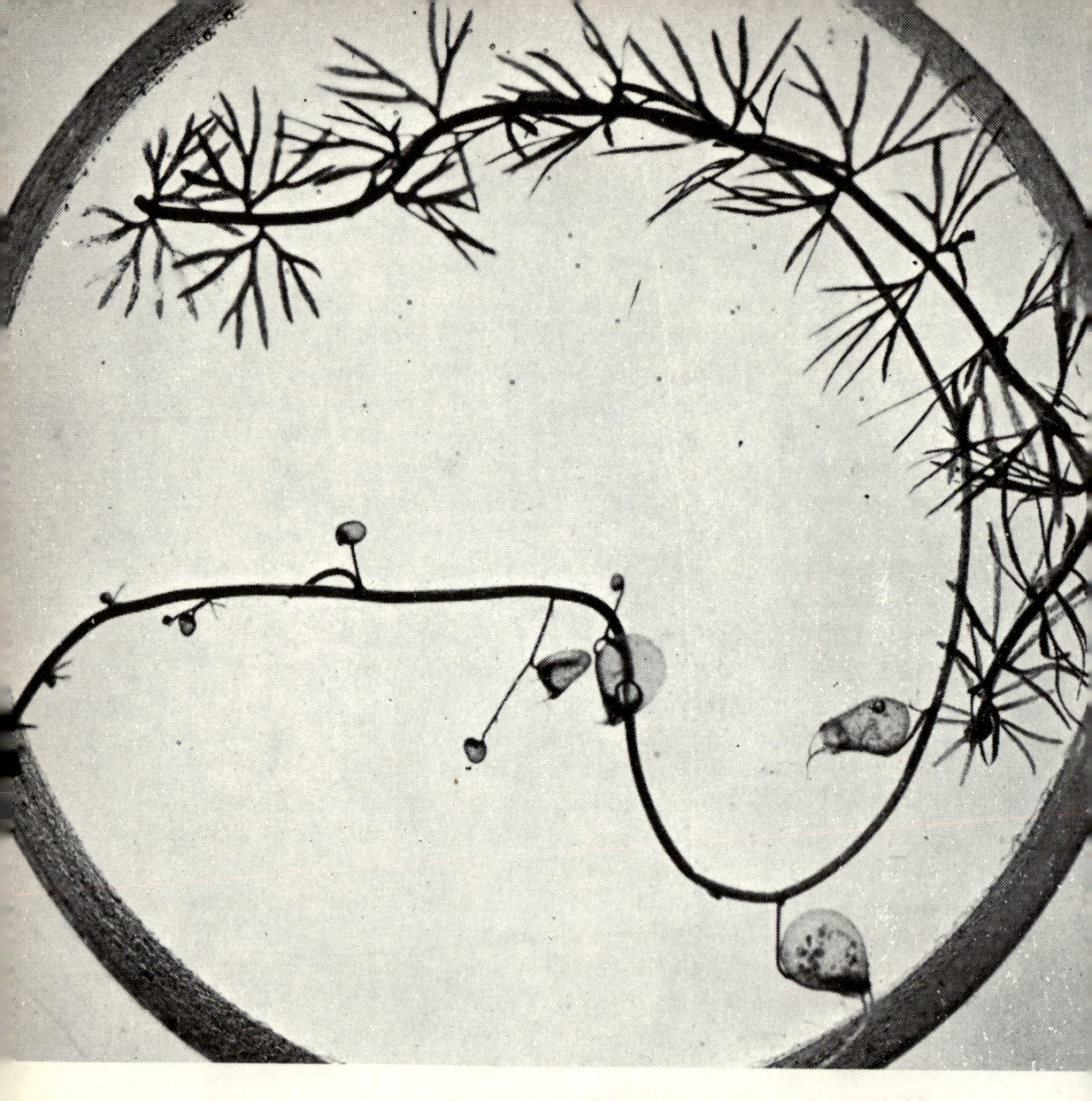

Bladderwort traps shown at different stages of maturity and magnified five times

seen with the naked eye, live in this plant debris, and some of them are trapped in the bladders. Each bladder is very small, tinier than the letter "o" on this page.

Some bladderworts neither grow beneath the soil nor in water. Instead, they grow on another plant, such as a wet stalk of a moss. However, they are not parasites, but *epiphytes*, meaning they do not depend on their host for food. The moss on which the bladderwort grows is only a roosting or growing place and does not provide food for the bladderwort. Food for this particular meat-eating plant comes from small creatures and organisms that live, drift, and fly about in the air. Water is also taken in from the air by this strange plant.

Swimming around where the water bladderworts grow are tiny organisms, from one-celled protozoa to larger animals, including insect larvae. At rest, each bladder is slightly compressed, somewhat like a rubber ball that has some of the air squeezed out of it. A small creature swimming around in the water touches some hairs that grow near the opening of the bladder, causing the bladder to inflate to its normal size. As it does so, it sucks in more water, and anything swimming in it. Usually, this is the creature that has touched the hairs originally. A valve closes around the opening, making it impossible for the animal to get out. The small creature dies and is digested and absorbed into the plant in a day or two, becoming a part of the food for the bladderwort. After most of the animal has been digested, water is forced out of the bladder, the bladder walls are sucked inward again, and the trap is set for the next victim.

Other water bladderworts besides the common, found in the wet spots of North America, are the purple bladderwort and the golden-horned bladderwort. The purple is so named because of its handsome purple flowers. The golden-horned bladderwort usually lives in a bog and has golden yellow flowers that bloom

Durchschnitt eines Schlauchblattes.

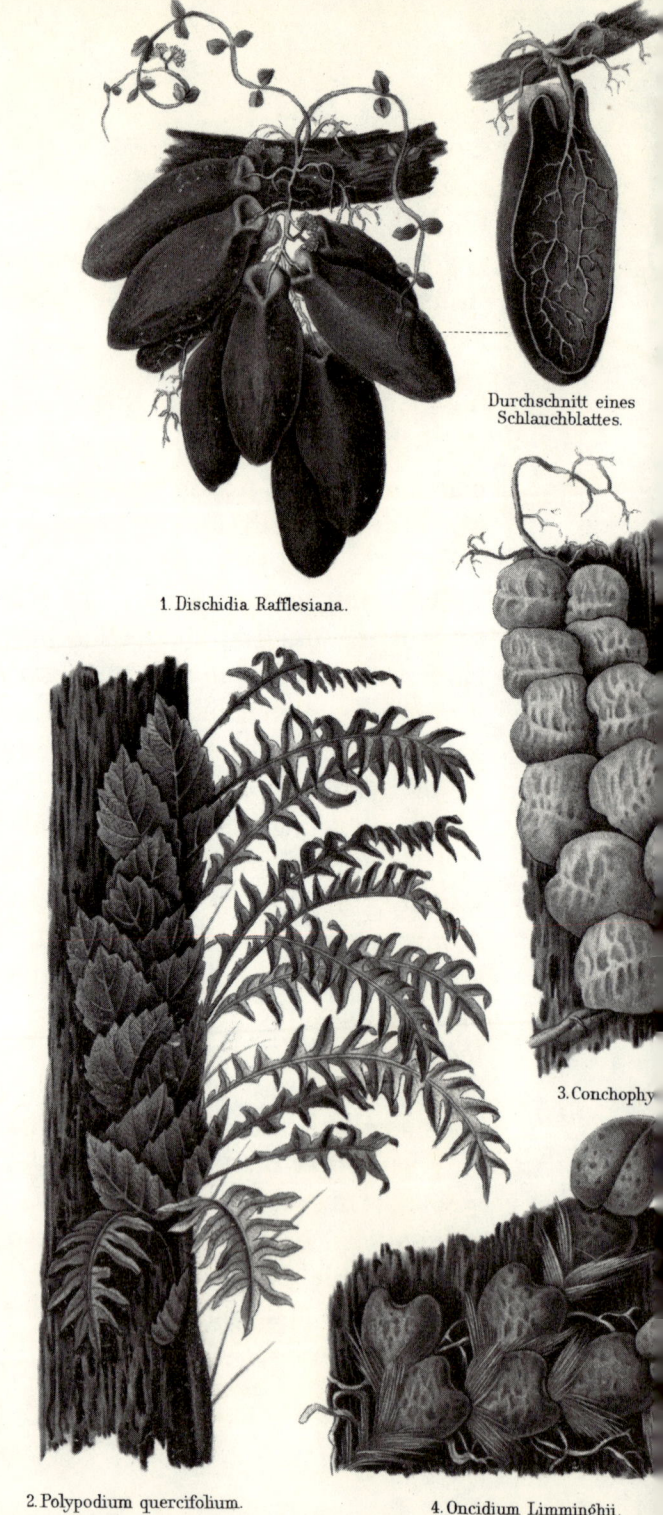

1. Dischidia Rafflesiana.

Different types of epiphytes, plants that do not depend on their hosts for food

2. Polypodium quercifolium.

3. Conchophy

4. Oncidium Limminghii.

5. Ficus religiosa.
6. Tillandsia bulbosa.
7. Platycerium grande.
8. Tillandsia usneoides.

A

B

C

in mid-summer. These yellow patches can be spotted from mountain peaks.

Another variety, the swollen bladderwort, has yellow flowers and a whorl of floats attached to its main stem. The whorl keeps the flowers from tipping during a breeze that may sweep across the surface of the pond where it grows.

Most of the bladders of bladderworts can be seen with a hand-held lens. A sample piece of stem can be snipped from a plant, but make sure that not too much is taken. When too much is snipped off, the plant may suffer damage.

Bladders were originally named by naturalists because it was first believed that the bladders were full of air and acted as floats to keep the plants from sinking to the bottom. Eventually, it was realized that these harmless-looking bladders were actually traps for capturing animals! Often when observing bladderworts with the aid of a lens, you may see the victim trapped inside. There may also be different kinds of plant debris trapped in the bladders.

Biologists and *botanists* are extremely interested in the small parts of the bladderwort. It is remarkable that such a tiny plant has such a complex system for catching food. Therefore, scientists study the bladderwort, its bladder, and traps in great detail. They want to learn as much as they can about how this strange plant functions.

The bladderwort:
A shows the entire plant;
B is a section through a bladder
with trapped small animals;
and C is a leaf showing
position of bladders

SUNDEWS

Anyone walking through bogs or marshes filled with the small carnivorous plants called sundews might think they resembled a field of sparkling gems. The common name of sundews comes from the glistening droplets at the ends of the hairs that cover the leaves and often the stem of the plant. There are ninety kinds of sundews, with variously shaped leaves and other differences, but they all have dazzling, sticky droplets on the hairs. These hairs make up the insect trap.

Sundews range in size from two to three inches across up to a foot or more. They grow in moist places across North America (except in the far northern climates), in parts of lower Canada, from southern New England to Florida and across to California, in European countries, Africa, Australia, and New Zealand.

Sundews have small leaves that are rounded in most, but not all, types. The leaves are green in color with a reddish-pink tinge. The plant grows horizontally, hugging the ground. The liquid on the hairs is clear and sticky. It gives off a sweet odor that attracts insects.

As insects fly above the bogs, heaths, or marshes, or crawl across the moss in search of food, they may become attracted to the sundew because of the glistening drops. When close to the plant, the insect is then attracted by the sweet odor.

The outside hairs, or tentacles, on each leaf of a sundew are longer than the tentacles in the center. When an insect, such as

An engraving of a round-leaved sundew

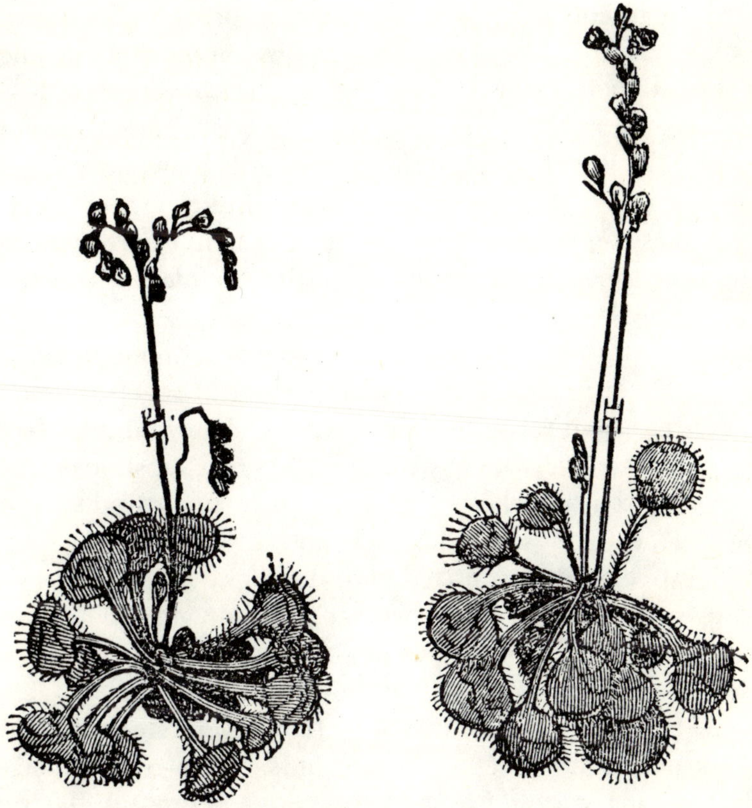

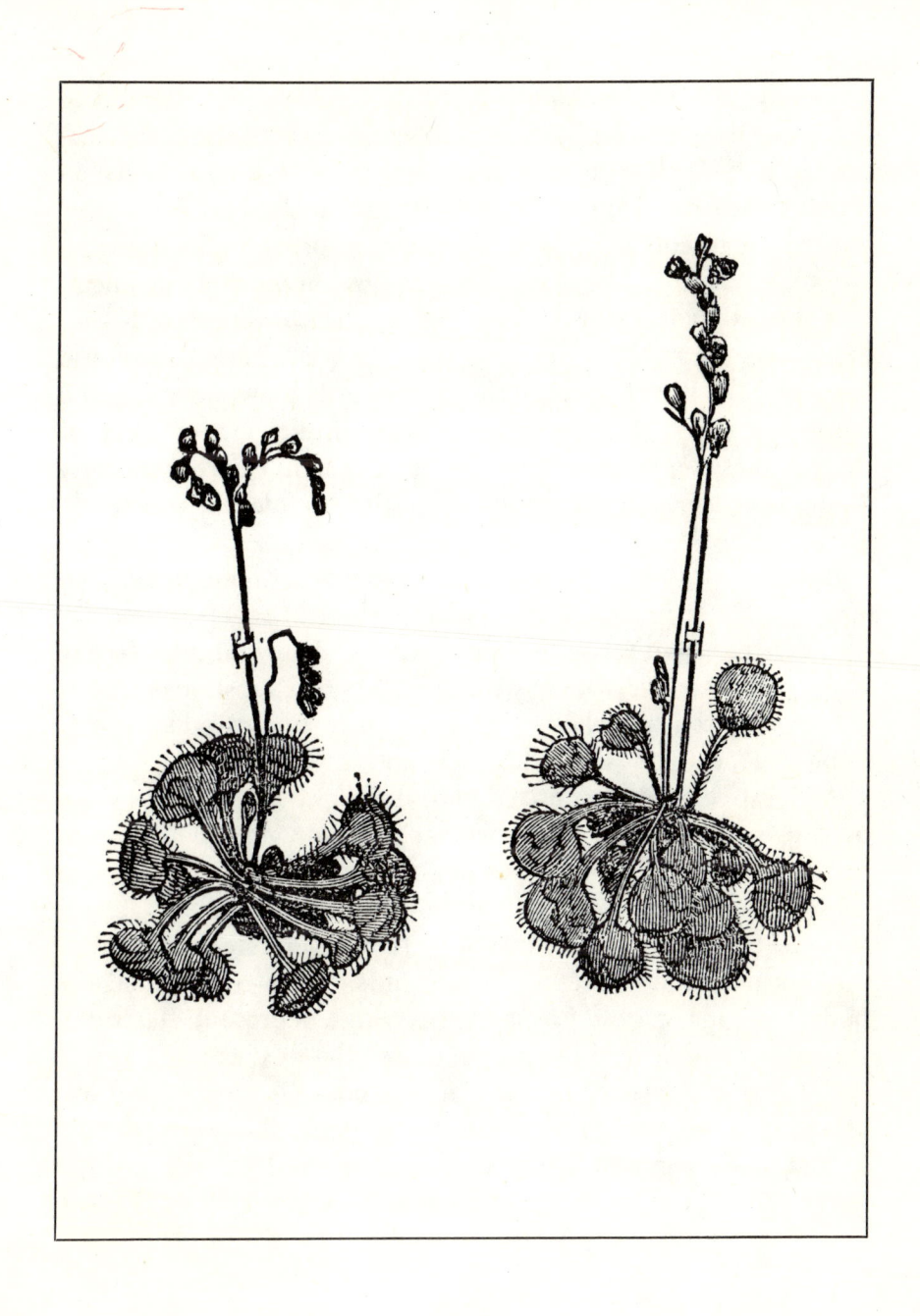

a deer fly, ant, or small moth lands on the leaf, it is trapped in the sticky droplets. As the insect struggles to free itself, the surrounding tentacles on the same leaf (those not actually in contact with the trapped animal) are stimulated to bend over toward the insect. They hold the victim in place, making escape almost impossible. The plant may also give off more sticky juices.

It takes four to eight days for the sundew to digest an insect, depending on its size. After this time the plant unfolds and the carcass of the insect drops to the ground or is blown away by the wind. The leaf is then ready to capture another victim. Each of these leaves can catch an insect, digest it, and open again only about three times. Then the leaf dies. But the plant is constantly growing new leaves, so the trapping continues.

If a small stone or a piece of sand or any other nonliving object lands on the tentacles of the sundew, the leaf reacts much differently than it does with a live animal. The tentacles barely move and do not secrete any additional sticky substance. One is almost tempted to believe that each plant knows when something is good to eat and when it is not.

A digestive solution is mixed with the sticky substance or gum. Scientists believe that the substance is an enzyme that helps to break down the animal's protein. The smaller the insect the shorter time needed to digest it, but for a good-sized fly it may take the full eight days for digestion to be complete. After digestion nothing is left except the hard outside skeleton, called the body case, made of *chitin*, and the wings if present. Then the leaf opens and secretes more, ready for the next catch.

Sundew plants vary in size. In Australia *Drosera binata* has leaves from six inches to about a foot long. They are deeply forked, each segment being very narrow, and having a slight curve toward the tip. Each leaf has a thick covering of crimson hairs, with a gland at the tip of each hair, or tentacle, about the size of a pinhead. The glands are covered with a sticky fluid that makes them shine like other sundews. Growing from the center of the plant is a tall stalk with a white flower streaked with yellow veins.

The sundew with the most brilliant flower is *Drosera cistiflora* from South Africa. It has brilliant scarlet blossoms about two inches across on top of a foot-long stem.

Leaves grow almost erect from another sundew called *Drosera capensis*. They are narrow, long, and thickly covered with slender, sparkling hairs that have sticky glands at the tips. Several of the deep pink, inch-wide flowers of the capensis grow on a single stem.

Drosophyllum lusitanicum, relatives of the Drosera, are found in Spain and Portugal. They have short woody stems from which the sparkling leaves grow much like other sundews. The Drosophyllum has yellow flowers, and, unlike the Droseras, its stem also has green leaves.

When the Portuguese and Spanish sundews snare an insect, their hairs cannot bend toward the captured prey. But the plant is still very successful in capturing flies and other insects. The Portuguese dig up the plants and take them home to use as fly catchers. These plants do well in houses. Unlike most carnivorous plants, Drosophyllum grow in hot and dry areas, such as the seashore or inland rocky places, under much drier conditions than many other meat-eaters.

A sundew called Byblis grows in western Australia. It is similar to the Portuguese sundew in that it does not roll its leaves toward the captured insect. It has a sticky gum or fluid to hold the prey in place. Strangely, a little bug can and does live among the tentacles of Byblis without any apparent harm. The bug

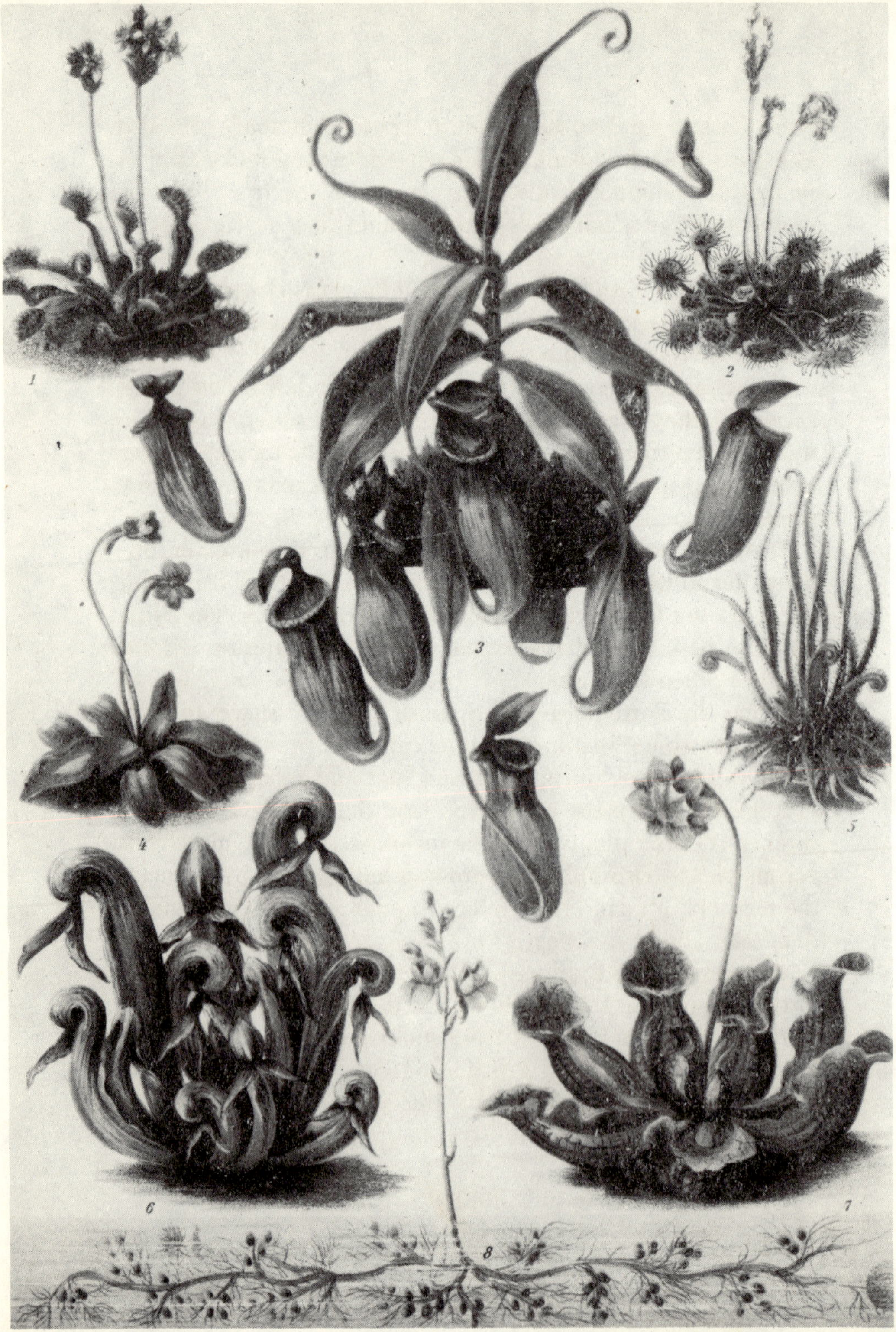
1
2
3
4
5
6
7
8

never becomes stuck to the leaves or digested. Actually, the bug steals food from the plant, helping itself to the flies, moths, or ants that are caught.

Some scientists, intrigued by the sundew and its digestion of insects, have performed experiments with it. Fruit flies were fed food containing a radioactive tracer. This was done by raising the very tiny flies in covered dishes in which sugar, yeast solutions, and some radiophosphorous had been placed. The object was to follow the path of the radioactive material from the insect into the sundew. These flies were eventually killed, washed to remove any surface radioactivity, and placed on the end of a single tentacle gland of the sundew. Several hours or days were allowed in various experiments for each plant to digest its fly. At varying intervals, flies were removed and the path of the absorbed nutrients followed by checking the roots, leaves, and flowers. It was learned that the plant absorbed the nutrients from the fly, including the radioactivity, very rapidly. During the first day food and radioactivity were absorbed into the leaf where the fly was placed. The material then moved quickly throughout the entire plant, with the greatest concentration occurring in the most actively growing areas. Through other experiments the scientists concluded that the plant does indeed obtain nutrients from its captured insects, and the absorption takes place through the tentacles or bristly hairs.

The Drosophyllum lusitanicum, *a sundew, is shown to the right in this picture, #5. (Also shown is the Venus Flytrap, #1; a sundew, #2; a pitcher plant, #3; a butterwort, #4; another pitcher plant, #6; the common pitcher plant, #7; and a bladderwort, #8.)*

Charles Darwin (1809–82), the great English scientist, conducted many studies with insect-eating plants. He placed such things as particles of glass and sand on the leaves of sundews and found that there was no active bending over of the tentacles on the particles and there was no secretion of additional sticky fluid or release of digestive juices. In some way that Darwin was unable to discover, the plant only released its juices when food was present.

When Darwin used bits of egg and meat, surrounding tentacles made contact with the food and released digestive juices. The plant reacted in the same way as if a live insect had been trapped. It reacted to chemicals in food whether the food was alive or dead.

THE VENUS FLYTRAP

After careful study and numerous experiments, Charles Darwin called the Venus Flytrap one of the most wonderful plants in the world. It grows in the wild only in North Carolina and in some parts of South Carolina in the United States. It is equipped to snap the two halves of its leaves shut like a steel animal trap. This meat-eater is a relative of the sundew.

The leaves of the flytrap grow in a rosette pattern. Most of the leaf is green, except for a reddish color on the inside of both halves. The first leaves lie along the ground with some of the ends curling upward. Each leaf is shaped like a ping-pong paddle, with the broad part hinged in the middle. Along the edge of each leaf are eighteen teeth, sometimes called spikes or spines. Nectar is also made along the outer edge of the leaf. It attracts insects. When an insect alights on the leaf, it eventually touches any of three sensitive hairs that grow in a triangle on the inner surface of each half. If the hairs are touched twice in succession, the leaf will snap shut in less than a second, trapping the insect. Spikes on each leaf half come together and mesh, allowing the halves to close very tightly. In so doing, the insect inside is squeezed against the digestive glands on the surface of the leaf. Once the leaf closes, it acts much like a stomach. The insect is digested. After several days the insect's body is softened and all that remains is the hard chitinous jacket.

The upper surface of each leaf-half, sometimes called a lobe, is thickly covered with small, purple glands. These glands can secrete digestive juices as well as absorb the digested insect. If the insect struggles, the plant only closes its leaves tighter and

*Leaves of the
Venus Flytrap*

secretes more juice. If a large insect is trapped, it may be able to crawl out of the leaf provided it is strong enough.

Part of the digestive juice, probably acid, in the leaves kills any bacteria that may be growing on the victim. Bacteria may arise because the insect remains inside the leaf many days. The liquid also keeps the insect from decaying or rotting while it is being absorbed, and instead of part of the insect decaying, most of it becomes available to the plant. As with other insect-eating plants, Venus Flytraps may derive nitrogen from insects.

When a leaf is activated it closes very quickly but not always with success. The leaves do catch many insects, but some escape. When the leaves close over something that is not suitable food, they do not squeeze shut but soon reopen. In twenty-four hours the leaves will be open almost all the way. However, it takes up to two days for the leaves to open fully once they have closed.

If an insect crawls into a leaf while it is reopening, the leaf can close just as rapidly as a fully open leaf. When a leaf does catch an insect, it usually remains closed for a long time, twenty or thirty days in some cases. After remaining closed for so many days, it reopens slowly and it may never close again. If the leaves do not reopen, they wither and die.

The plants bloom in early June with white flowers that show above the grasses of the bogs. Joining them are the blossoms of blueberries, pyxie flowers, and blazing stars. The small white flowers grow from a slender white stem about six to eight inches high. The plant does not have very many roots.

Darwin experimented a great deal with the Venus Flytrap. Among other things, he learned that wind and rain had no effect on the leaves. In experiments he dropped small amounts of water on the leaves, and they did not close. He poured a great deal of water on the leaves, and they still did not close. He even added sugar to the water, and the leaf remained open. He blew on the leaf very strongly with a pointed tube, and the leaf did not move. From all this Darwin deduced that rain and wind had no effect, even a heavy gale. If the leaf was placed completely in water it sometimes closed, but, as Darwin noted, this was an unnatural condition.

According to Darwin, when a leaf first closes a small opening remains between the spikes, allowing tiny insects to escape before the leaf closes entirely. The insects escape through the small openings left by the crisscrossing of the spikes. They are similar to a fishnet that allows small fish to escape but keeps larger fish trapped. It might be concluded that it is beneficial to the plant to allow very small insects to get away. Otherwise the Venus Flytrap would be closed for many days over an insect that would not give much nutrition.

Just what makes the Venus Flytrap close its leaves has not been discovered as yet. For at least 200 years scientists have studied the plant, but have not come up with the answer. However, the studies have resulted in some known facts.

The Venus Flytrap does not have any kind of nervous system as an animal does. There are no muscles or tendons as there are in functioning animals. One theory is that a shift of fluids makes the lobe shut quickly and tightly. Scientists have detected an electrical current in the leaves, and this may have something to do with the closing of the leaf.

The Venus Flytrap was first described by a British colonial governor in North Carolina in 1760. He wrote of "a very curious

*Details of
Venus Flytrap leaves*

A

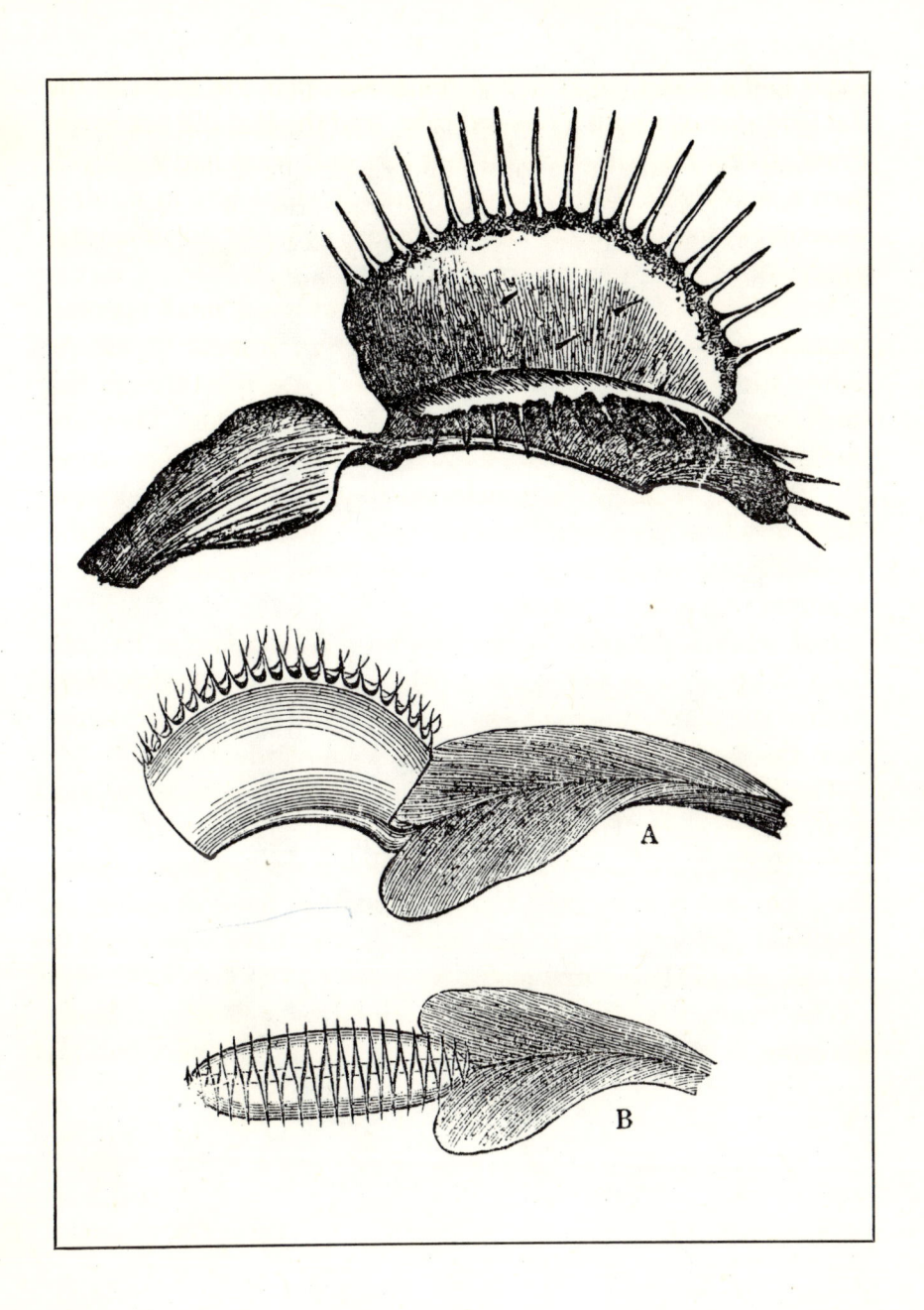

B

and unknown plant that was able to close like a spring trap. It also confined an insect or anything that fell between the leaves."

A short time later a botanist from England collected some of the plants from North Carolina. He took them to London for observation. A merchant whose hobby was studying plants named the Venus Flytrap *Dionaea muscipula*.

BUTTERWORTS

Butterworts are small plants often seen by hikers on alpine trails or across the moors. They grow in marshes and bogs, and because of their small purple flowers, they are sometimes referred to as bog violets.

These insect-eaters are found in cool climates across the Northern Hemisphere. They grow in the United States, Canada, Europe, the U.S.S.R., and other countries of similar latitude around the world. There are as many as thirty different kinds of butterworts, and many of them look alike. The average diameter of the plant is two inches, but some grow to be five inches across.

Each plant has about eight leaves some one and a half inches long and three-quarters of an inch wide that grow in a rosette. They are terrestial plants, meaning that they do not grow in water but in sphagnum moss, on wet or mossy logs, or on wet sand. Pitcher plants and sundews grow in similar conditions.

In sunlight the leaves of the butterwort glisten. They are yellowish-green in color, and the outside, bladelike edges curl in. Because of the yellow color of the leaves and their greasy appearance, they have been tagged with the common name of butterwort. The plant's Latin name is *Pinguicula,* from the Latin *pinguis*, which means "fat."

A sticky fluid is secreted by the leaves. This fluid comes from two kinds of glands. One type is stalked and produces a gluey substance; the other, without stalks, produces a digestive enzyme and absorbs the material that is digested.

Each leaf also gives off an odor that is musty or damp smelling. The sticky substance and smell attract midges, gnats, and other small creatures. When one of these insects is trapped, it

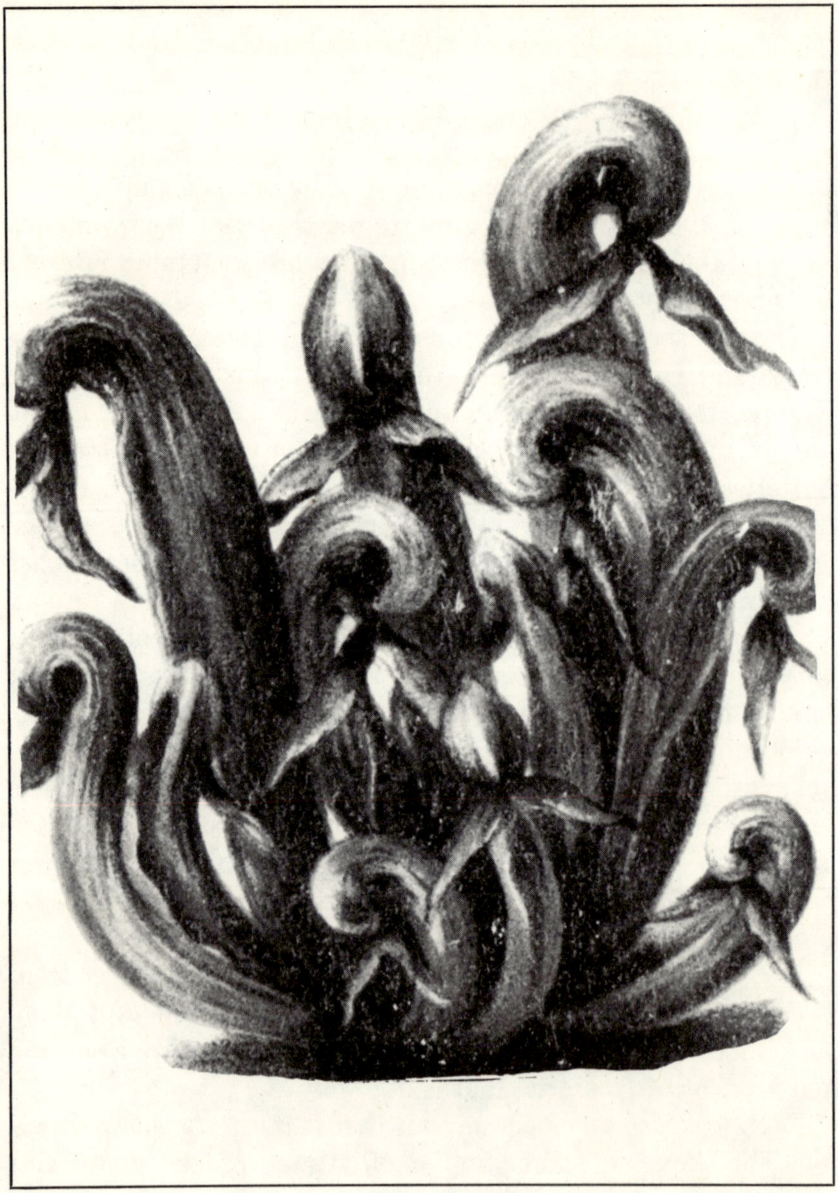

struggles, causing the leaf edges to roll in and engulf the animal. The insect is pushed toward the middle, and the glands produce the digestive juices.

Unlike the Venus Flytrap, the curling of the butterwort leaf has nothing to do with the capture of the insect. It only curls in to bring the insect into contact with more digestive glands.

Once trapped, the insect breaks down, and the protein is absorbed by the glands into the leaves and to other parts of the butterwort. After two or three days the leaf will reopen and reveal what is left of the insect.

The butterwort is able to catch only very tiny insects. Big ones are able to crawl away. When nonliving material, such as a grain of sand, lands on the leaf, the glands secrete little if any fluid. The plant may absorb certain substances from seeds or pollen and anything else with food value whether it is animal or vegetable.

One kind of butterwort, *Pinguicula caudata,* with striking large flowers of pink or violet, grows in Mexico. The flowers may be two inches across, and orchid-growers raise them in warm orchid houses. Despite their beauty, the plant growers have a single purpose for raising them. The butterworts help to rid orchid houses of midges, an orchid pest.

Years ago people used the butterwort for medicinal purposes. The Swiss believed that the greasy-looking leaves had health-giving properties. They rubbed them on open wounds of animals, especially cattle. After treatment the sores healed, and the practice still continues today in certain Swiss mountain communities.

The butterwort known as Pinguicula vulgaris

Farmers also use the butterwort to make cheese. In earlier times it was learned that the leaves of the butterwort, when placed in milk, caused the milk to curdle. Dairymen in Sweden and Denmark, as well as in other parts of northern Europe and the U.S.S.R., have been using butterwort leaves for cheese-making for several centuries. They cut the leaves from the butterwort plant and place them in kegs or cans filled with milk. The acid in the leaves causes the milk to curdle. After the curdling is complete, the milk is made into cheese. Of all the carnivorous plants, the butterwort is the only one known to be so useful to man.

ANIMAL-CAPTURING FUNGI

Fungi are plants, but they differ a great deal from green plants. They have no stems, no leaves, and no chlorophyll. Since they lack chlorophyll, fungi are not able, as are green plants, to make their own food. They are dependent on getting food from organic matter such as dead logs or the decaying matter found in soil. In many cases, they are parasites.

These plants range in size from puff balls that may be as big as a bushel basket to slime molds about the size of a pinhead. They also come in many colors, from reds, oranges, and yellows, to greens, purples, and even some browns and blacks.

Besides not having leaves or stems, fungi do not have roots. Instead, they have many fine threads that are usually found belowground. These fine threads are called *hyphae*. A mass of hyphae is often called a *mycelium*. The part of the mycelium that is seen, for example, aboveground or on the bark of a tree is the fruiting body. A mushroom is a fruiting body of a particular fungus.

Fungi do not produce seeds. Instead, they have spores, formed in the fruiting body.

Molds, mushrooms, puffballs, and cup fungi are common examples of fungi. Those that live off living plants are called parasites, and those that live off dead plants or animals are called *saprophytes*. Some fungi are both. Most of the different kinds of fungi do not digest their own food in their own plant bodies. They secrete enzymes onto or into the plant or animal material on which they are living, and in that way obtain nutrients.

Some fungi feed on tiny worms and protozoa that live on plants or in the soil. These protozoa and worms are microscopic,

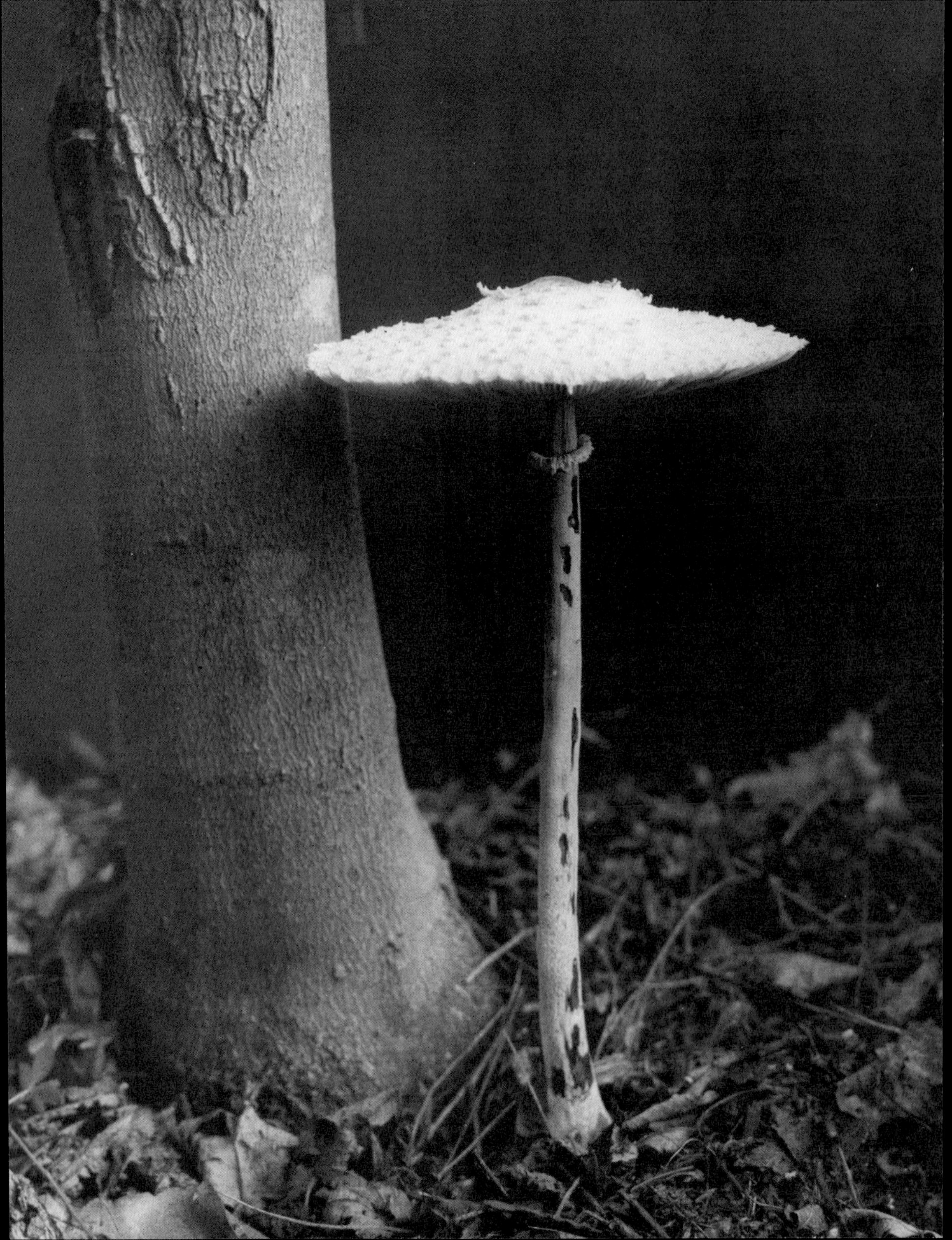

Above: a puffball
Left: a mushroom growth beside a tree

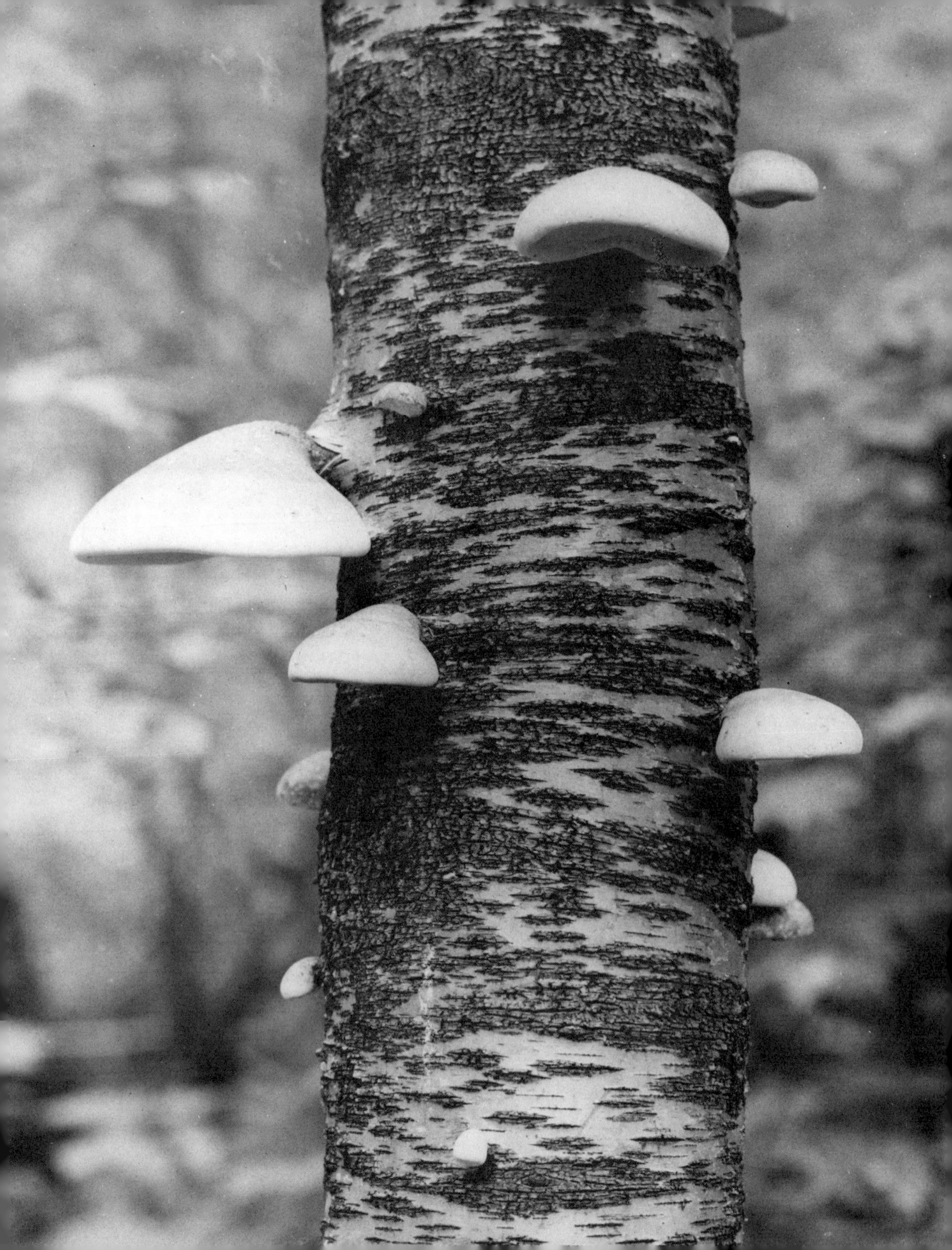

meaning that they cannot be seen without the aid of a microscope. There are about twenty different kinds of animal-eating fungi.

These fungus plants "lasso" their food. Some of the short branches on the hyphae bend over and form a loop, which forms the trap.

Scientists have conducted studies with such fungi, and they were not able to fool the tiny plant anymore than they could fool the larger carnivorous plants. Some scientists tried to trigger the loops by inserting small wires into them. The fungus loops did not tighten around the wires. The scientists even warmed the wires to see if temperature was important, but the loops remained cocked waiting for live prey.

Some kinds of fungus can catch their quarry with glue. They have hundreds of sticky parts that trap a tiny animal. Then the fungus feeds on it. In one case, after an animal is caught, it is covered by new fungus threads. When the victim is completely covered, the fungus begins to feed on it.

Some animal-capturing fungi have sticky traps complete with poison. After the animal is clamped onto the sticky substance, it is jabbed with a stinger. The stinger contains poison that kills the animal, and the fungus then feeds on the victim.

Mushroom growths on tree trunk

🙢 GLOSSARY 🙠

Algae — one- or many-celled plants, mostly aquatic, containing chlorophyll.
Annual — plant that completes its life cycle in one season.
Biologist — scientist who studies living organisms.
Bladderwort — carnivorous plant with leaves, twisting stems, and bladders that trap prey.
Botanist — scientist who studies plant life.
Butterwort — carnivorous plant with yellowish, greasy-looking leaves that captures prey with a sticky substance.
Carnivorous — flesh-eating; when referring to carnivorous plants it means that plant is capable of digesting and absorbing insects; does not actively catch and eat them.
Chitin — material that makes up the hard body case of an insect.
Chlorophyll — green pigment in plants.
Epiphyte — plant that grows on another plant for support only.
Fungi — plants without roots, stems, or leaves that lack chlorophyll but have fine threads usually found belowground.
Hyphae — fine threads making up the fungus.
Larva — juvenile form of an insect.
Mycelium — mass of hyphae.
Parasite — organism that lives in or on another organism (host), from which it obtains foods.
Perennial — plant that continues its growth from year to year.
Pitcher plant — carnivorous plant with leaves shaped like pitchers.
Saprophyte — organism that lives off dead plant or animal matter.
Sepal — usually green outer covering of some flowers.

Sundew — carnivorous plant with glistening droplets at the end of the hairs that cover the leaves and often the stem.

Terrarium — enclosed glass case for growing carnivorous or other plants.

Venus Flytrap — carnivorous plant that snaps its leaf halves shut like a steel animal trap when insect alights on leaf and touches sensitive hairs.

❧ FOR FURTHER READING ❧

Anyone who enjoyed this book on carnivorous plants may also be interested in the following:

Budlong, Ware & Fleitzer, Mark H. *Experimenting with Seeds and Plants.* New York: G. P. Putnam's, 1970.
Cavanna, Betty. *The First Book of Wildflowers.* New York: Franklin Watts, 1960.
Cobb, Vicki. *Cells: The Basic Structure of Life.* New York: Franklin Watts, 1970.
Dickinson, Alice. *The First Book of Plants.* New York: Franklin Watts, 1963.
Fenton, D. X. *Gardening... Naturally.* New York: Franklin Watts, 1973.
Hoke, John. *Terrariums.* New York: Franklin Watts, 1972.
Rahn, Joan E. *How Plants Travel.* New York: Atheneum, 1972.
Tribe, Ian. *Plant Kingdom.* New York: Bantam Books, 1971.

INDEX

Algae, 2
Annuals, 16

Bladderworts, 9, 25-31
    bladders, 25, 27, 31
    locations, 25, 27
    types of 25, 27, 31
    Utricularia, 25
Butterworts, 9, 45-48
    insect trap, 45, 47
    locations, 45
    Pinguicula caudata, 47

Carnivorous plants, definition of, 6-13
    See also Bladderworts, Butterworts, Pitcher plants, Sundews, *and* Venus Flytraps.
Carnivorous plants in the home, 10, 13
Carnivorous plants in the wild, 10
Catesby's Pitcher Plant. *See* Pitcher plants.
Chitin, 14, 34
Chlorophyll in plants, 2, 9
Cobra plant. *See* Pitcher plants.

Darwin, Charles, 38, 39, 41, 42

Enzymes, 16, 34
Epiphytes, 27
Experiments with carnivorous plants, 37, 38, 39, 41, 42
Extinction, threat of, 9-10

Flesh-eating plants. *See* Carnivorous plants, definition of.
Flowers. *See* Pollination.
Fungi, animal-capturing, 49-53

Hyphae, 49

Insect-eating plants. *See* Carnivorous plants, definition of.
Insects as pollinators, 2-4
Insects on plants, 1

Larvae, 21, 27
Leaves. *See* Parts of a plant.

Man-eating plants. *See* Carnivorous plants, definition of.
Mushrooms, 2
    See also Fungi.
Mycelium, 49

(59)

Parasites, 2, 27
Parrot pitcher plant. *See* Pitcher plants.
Parts of a plant, 5
Perennials, 16
Pitcher plants, 8, 13, 14-24
    Chrysamphora californica, 21
    insect trap, 16-17, 21, 22
    leaves, 16, 17
    locations, 14, 16, 17, 21, 22, 24
    Nepenthes rajah, 24
    Nepenthes tentaculata, 22
    Sarracenia flava, 21
    Sarracenia psittacina, 21
    Sarracenia purpurea, 17
    Sarracenia rubra, 21
Plant uses, 1
Plants as decoration, 1
Plants as food, 1
Plants, location of, 1
Plants, types of, 1-2
Pollination, 2-4
Protozoa, 27

Reproduction in plants, 2

Roots. *See* Parts of a plant.

Saprophytes, 49
Seaweeds, 2
Seeds, growing carnivorous plants from, 13
    See also Pollination.
Sepals, 16
Sequoia trees, 2
Stem. *See* Parts of a plant.
Sundews, 9, 13, 32-36
    Byblis, 35, 37
    Drosera binata, 34-35
    Drosera cistiflora, 35
    Drosera copenis, 35
    Drosophyllum lusitanicum, 35
    insect traps, 32, 34
    locations, 32, 34, 35

Terrariums, 10, 13
Trees, 2

Venus Flytraps, 9, 10, 13, 39-44, 47
    Dionaea muscipula, 44
    insect trap, 39, 41, 42
    locations, 39

ABOUT THE AUTHOR

Born and educated in Massachusetts (he majored in education and science), John Waters now makes his home in Northeast Harbor, Maine, and Chatham, Massachusetts. A former newspaper writer and elementary schoolteacher, the author has written a number of books on various scientific subjects, including eels, green turtles, mammals that live in the sea, oceanographers, sharks, and the New England shore. He is married to an elementary science specialist, and they have four children. This is his first book for Franklin Watts.

J583   295121
Waters
Carnivorous plants.

Johnson Free Public Library
Hackensack, New Jersey

3 9123 00154406 8
HACKENSACK-JOHNSON LIBRARY
a39123001544068b